Tabulated area = probability

Standard normal (z) curve

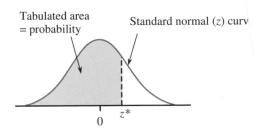

0   z^*

z^*	.00	.01	.02	.03	.04	.05	.06	.07	.08	.09
0.0	.5000	.5040	.5080	.5120	.5160	.5199	.5239	.5279	.5319	.5359
0.1	.5398	.5438	.5478	.5517	.5557	.5596	.5636	.5675	.5714	.5753
0.2	.5793	.5832	.5871	.5910	.5948	.5987	.6026	.6064	.6103	.6141
0.3	.6179	.6217	.6255	.6293	.6331	.6368	.6406	.6443	.6480	.6517
0.4	.6554	.6591	.6628	.6664	.6700	.6736	.6772	.6808	.6844	.6879
0.5	.6915	.6950	.6985	.7019	.7054	.7088	.7123	.7157	.7190	.7224
0.6	.7257	.7291	.7324	.7357	.7389	.7422	.7454	.7486	.7517	.7549
0.7	.7580	.7611	.7642	.7673	.7704	.7734	.7764	.7794	.7823	.7852
0.8	.7881	.7910	.7939	.7967	.7995	.8023	.8051	.8078	.8106	.8133
0.9	.8159	.8186	.8212	.8238	.8264	.8289	.8315	.8340	.8365	.8389
1.0	.8413	.8438	.8461	.8485	.8508	.8531	.8554	.8577	.8599	.8621
1.1	.8643	.8665	.8686	.8708	.8729	.8749	.8770	.8790	.8810	.8830
1.2	.8849	.8869	.8888	.8907	.8925	.8944	.8962	.8980	.8997	.9015
1.3	.9032	.9049	.9066	.9082	.9099	.9115	.9131	.9147	.9162	.9177
1.4	.9192	.9207	.9222	.9236	.9251	.9265	.9279	.9292	.9306	.9319
1.5	.9332	.9345	.9357	.9370	.9382	.9394	.9406	.9418	.9429	.9441
1.6	.9452	.9463	.9474	.9484	.9495	.9505	.9515	.9525	.9535	.9545
1.7	.9554	.9564	.9573	.9582	.9591	.9599	.9608	.9616	.9625	.9633
1.8	.9641	.9649	.9656	.9664	.9671	.9678	.9686	.9693	.9699	.9706
1.9	.9713	.9719	.9726	.9732	.9738	.9744	.9750	.9756	.9761	.9767
2.0	.9772	.9778	.9783	.9788	.9793	.9798	.9803	.9808	.9812	.9817
2.1	.9821	.9826	.9830	.9834	.9838	.9842	.9846	.9850	.9854	.9857
2.2	.9861	.9864	.9868	.9871	.9875	.9878	.9881	.9884	.9887	.9890
2.3	.9893	.9896	.9898	.9901	.9904	.9906	.9909	.9911	.9913	.9916
2.4	.9918	.9920	.9922	.9925	.9927	.9929	.9931	.9932	.9934	.9936
2.5	.9938	.9940	.9941	.9943	.9945	.9946	.9948	.9949	.9951	.9952
2.6	.9953	.9955	.9956	.9957	.9959	.9960	.9961	.9962	.9963	.9964
2.7	.9965	.9966	.9967	.9968	.9969	.9970	.9971	.9972	.9973	.9974
2.8	.9974	.9975	.9976	.9977	.9977	.9978	.9979	.9979	.9980	.9981
2.9	.9981	.9982	.9982	.9983	.9984	.9984	.9985	.9985	.9986	.9986
3.0	.9987	.9987	.9987	.9988	.9988	.9989	.9989	.9989	.9990	.9990
3.1	.9990	.9991	.9991	.9991	.9992	.9992	.9992	.9992	.9993	.9993
3.2	.9993	.9993	.9994	.9994	.9994	.9994	.9994	.9995	.9995	.9995
3.3	.9995	.9995	.9995	.9996	.9996	.9996	.9996	.9996	.9996	.9997
3.4	.9997	.9997	.9997	.9997	.9997	.9997	.9997	.9997	.9997	.9998
3.5	.9998	.9998	.9998	.9998	.9998	.9998	.9998	.9998	.9998	.9998
3.6	.9998	.9998	.9999	.9999	.9999	.9999	.9999	.9999	.9999	.9999
3.7	.9999	.9999	.9999	.9999	.9999	.9999	.9999	.9999	.9999	.9999
3.8	.9999	.9999	.9999	.9999	.9999	.9999	.9999	.9999	.9999	1.0000

Statistics: The Exploration and Analysis of Data

Statistics

*The Exploration
and Analysis of Data*

Fourth Edition

Jay Devore
*California Polytechnic State University
San Luis Obispo*

Roxy Peck
*California Polytechnic State University
San Luis Obispo*

DUXBURY
™
THOMSON LEARNING

Australia • Canada • Mexico • Singapore • Spain • United Kingdom • United States

DUXBURY

THOMSON LEARNING

Sponsoring Editor: *Carolyn Crockett*
Assistant Editor: *Seema Atwal*
Marketing Team: *Tom Ziolkowski/
  Samantha Cabaluna*
Editorial Assistant: *Ann Day*
Production Editor: *Tessa Avila*
Production Service: *Susan L. Reiland*
Manuscript Editor: *Christine Levesque*

Permissions Editor: *Fiorella Ljunggren*
Interior Design: *Andrew Ogus*
Cover Design: *Cassandra Chu*
Interior Illustration: *Lori Heckelman*
Print Buyer: *Vena Dyer*
Typesetting: *G & S Typesetters*
Cover Printing: *Phoenix Color Corp.*
Printing and Binding: *R. R. Donnelley/Crawfordsville*

For more information about this or any other Duxbury product, contact:
DUXBURY
511 Forest Lodge Road
Pacific Grove, CA 93950 USA
www.duxbury.com
1-800-423-0563 (Thomson Learning Academic Resource Center)

Printed in United States of America

10 9 8 7 6 5 4 3 2 1

Library of Congress Cataloging-in-Publication Data
Devore, Jay L.
    Statistics: the exploration and analysis of data / Jay Devore, Roxy Peck.
      p. cm.
    Includes bibliographical references and index.
    ISBN 0-534-35867-5
    1. Statistics.   I. Title: Exploration and analysis of data.   II. Peck, Roxy.   III. Title.
QA276.D48 2001
519.5 — dc21

00-033736

To our families, colleagues, friends, and students,
who have given us so much support over the years.

Contents

Answers to Selected Odd-Numbered Exercises 695

Index 709

Preface

Statistics: The Exploration and Analysis of Data, 4th edition, is intended for use as a textbook in introductory statistics courses at two- and four-year colleges and universities. We believe that the following special features of our book distinguish it from other texts.

Features

A Traditional Structure with a Modern Flavor

The topics included in almost all introductory texts are here also. However, we have interwoven some new strands that reflect current and important developments in statistical analysis. These include coverage of sampling and experimental design, the role of graphical displays as an important component of data analysis, transformations, residual analysis, normal probability plots, and simulation. The organization gives instructors considerable flexibility in deciding which of these topics to include in a course.

The Use of Real Data and the Importance of Context

Many students are skeptical of the relevance and importance of statistics. Contrived problem situations and artificial data often reinforce this skepticism. A strategy that we have employed successfully to motivate students is to present examples and exercises that involve data extracted from journal articles, newspapers, and other published sources. Most examples and exercises in the book are of this nature; they cover a very wide range of disciplines and subject areas. These include, but are not limited to, health and fitness, consumer research, psychology and aging, environmental research, law and criminal justice, and entertainment.

Statistics is not about numbers; it is about data — numbers in context. It is the context that makes a problem meaningful and something worth considering. Examples and exercises with overly simple settings do not allow students to practice interpreting results in authentic situations or give students the experience necessary to be able to use statistical methods in real settings. We believe that the exercises and examples are a particular strength of this text, and we invite you to compare the problem scenarios with those in other statistics books.

Mathematical Level and Notational Simplicity

A good background in high-school algebra constitutes sufficient mathematical preparation for reading and understanding the material presented herein. We want students to focus on concepts without having to grapple unnecessarily with the manipulation of formulas and symbols. To achieve this, we have sometimes used words and phrases in addition to and in place of symbols, as shown here in material from pages 119 and 370. (See page 336 for another example.) For those who are apprehensive about their mathematical skills, we trust the verbal descriptions are not only faithful to the statistical concepts but also stepping-stones to the more precise mathematical descriptions.

EXAMPLE 4.17

Suppose that two graduating seniors, one marketing major and one accounting major, are comparing job offers. The accounting major has an offer for $35,000 per year, and the marketing student has one for $33,000 per year. Summary information about the distribution of offers follows:

Accounting: mean = 36,000   standard deviation = 1500

Marketing: mean = 32,500   standard deviation = 1000

Then,

$$\text{accounting } z \text{ score} = \frac{35,000 - 36,000}{1500} = -.67$$

(so $35,000 is .67 standard deviation below the mean), whereas

$$\text{marketing } z \text{ score} = \frac{33,000 - 32,500}{1000} = .5$$

Relative to the appropriate data sets, the marketing offer is actually more attractive than the accounting offer (though this may not offer much solace to the marketing major).

Summary of Large-Sample z Test for π

Null hypothesis: H_0: π = hypothesized value

Test statistic: $z = \dfrac{p - \text{hypothesized value}}{\sqrt{\dfrac{(\text{hypothesized value})(1 - \text{hypothesized value})}{n}}}$

Alternative hypothesis:

H_a: π > hypothesized value

H_a: π < hypothesized value

H_a: π ≠ hypothesized value

P-value:

Area under z curve to right of calculated z

Area under z curve to left of calculated z

(i) 2·(area to right of z) if z is positive

(ii) 2·(area to left of z) if z is negative

Assumptions: 1. p is the sample proportion from a *random sample.*

2. The *sample size is large.* This test can be used if n satisfies both $n(\text{hypothesized value}) \geq 10$ and $n(1 - \text{hypothesized value}) \geq 10$.

The Use of Technology

The computer has brought incredible statistical power to the desktop of every investigator. The wide availability of statistical computer packages such as MINITAB, S-Plus, JMP, and SPSS, and the graphical capabilities of the modern microcomputer have transformed both the teaching and learning of statistics. To highlight the role of the computer in contemporary statistics, we have included sample output throughout the book, such as that shown here from pages 169 and 386. In addition, numerous exercises contain data that can easily be analyzed by computer, though our exposition firmly avoids a presupposition that students have access to a particular statistical package.

FIGURE 5.21 MINITAB output for the data of Example 5.14

Regression Analysis

The regression equation is
Range of Motion = 108 + 0.871 Age

Predictor	Coef	StDev	T	P
Constant	107.58	11.12	9.67	0.000
Age	0.8710	0.4146	2.10	0.062

S = 10.42   R-Sq = 30.6%   R-Sq(adj) = 23.7%

Analysis of Variance

Source	DF	SS	MS	F	P
Regression	1	479.2	479.2	4.41	0.062
Residual Error	10	1085.7	108.6		
Total	11	1564.9			

SSResid (annotation pointing to 1085.7)

SSTo (annotation pointing to 1564.9)

9. Because the P-value $> \alpha$, we fail to reject H_0. There is not sufficient evidence to conclude that the mean time spent in personal use of company technology is greater than 75 minutes per day for this company.

MINITAB could also have been used to carry out the test, as shown in the accompanying output.

T-Test of the Mean
Test of mu = 75.00 vs mu > 75.00

Variable	n	Mean	StDev	SE Mean	t	P
Time	10	74.80	9.45	2.99	-0.07	0.53

Although we had to round the computed t value to $-.1$ to use Appendix Table IV, MINITAB was able to compute the P-value corresponding to the actual value of the test statistic.

Topic Coverage

Our book can be used in courses as short as one quarter or as long as one year in duration. Particularly in shorter courses, an instructor will need to be selective in deciding which topics to include and which to set aside. The book divides naturally into four major sections: collecting data and descriptive methods (Chapters 2–5), probability material (Chapters 6–8), the basic one- and two-sample inferential techniques (Chapters 9–11), and more advanced inferential methodology (Chapters 12–15). We have joined a growing number of books in including an early chapter (Chapter 5) on descriptive methods for bivariate numerical data. This early exposure raises questions and issues that should stimulate student interest in the subject; it is also advantageous for those teaching courses in which time constraints preclude covering advanced inferential material. However, this chapter can easily be postponed until the basics of inference have been covered, and then combined with Chapter 13 for a unified treatment of regression and correlation.

With the possible exception of Chapter 5, Chapters 1–10 should be covered in order. We anticipate that most instructors will then continue with the two-sample material of Chapter 11, although regression could be covered before either of these. Analysis of variance (Chapter 15), and/or categorical data analysis (Chapter 12) can be discussed prior to the regression material of Chapters 13–14. In addition to flexibility in the order in which chapters are covered, material in some sections can be skipped entirely or in part postponed. The authors would be happy to provide more detailed suggestions concerning coverage.

A Note on Probability

This book takes a brief and informal approach to probability, focusing on those concepts needed to understand the inferential methods covered in the later chapters. For those who prefer a more traditional approach to probability, the book *Introduction to Statistics and Data Analysis* by Roxy Peck, Chris Olsen, and Jay Devore (also published by Duxbury Press) may be a more appropriate choice. Except for the more formal treatment of probability and the inclusion of optional Graphing Calculator Explorations, it parallels the material in this text.

New in This Edition

There are a number of changes in the fourth edition, including

- A new chapter with expanded coverage of sampling and experimental design.
- A new section on interpreting the results of statistical analysis included in most chapters.
- Expanded coverage of descriptive and graphical methods, and an increased emphasis on the use of graphical display as a part of the data analysis process.
- Expanded coverage of transformations and nonlinear models.
- Decreased emphasis on inferential methods that assume a known population standard deviation.
- Many new examples and exercises that use data from current journals and newspapers.
- Reorganization of the later chapters to group chapters containing topics that are not always covered in a one-quarter or one-semester course.

Acknowledgments

Many people have made valuable contributions to the preparation of this book. Carolyn Crockett, our editor at Duxbury, has been tremendously supportive throughout, as has our wonderful production coordinator, Susan Reiland. Many others at Duxbury have been helpful in bringing this project to completion, and we thank them for their support and help. Steve Rein performed an accuracy check and made many good suggestions for improving our manuscript.

We are also grateful to the constructive suggestions that came from the following manuscript reviewers: Patricia M. Buchanan, Pennsylvania State University; Mary C. Christman, University of Maryland; Dale O. Everson, University of Idaho; Mark E. Glickman, Boston University; John Z. Imbrie, University of Virginia; Pamela Martin, Northeast Louisiana University; Michael J. Phelan, Chapman University; and Alan Polansky, Northern Illinois University.

And, as always, we thank our families, friends, and colleagues for their continued support.

Jay Devore
Roxy Peck

1

The Role of Statistics

INTRODUCTION

Statistical methods for summary and analysis provide investigators with powerful tools for making sense out of data. Statistical techniques are being employed with increasing frequency in business, medicine, agriculture, social sciences, natural sciences, and applied sciences such as engineering. The pervasiveness of statistical analyses in such diverse fields has led to increased recognition that statistical literacy — a familiarity with the goals and methods of statistics — is a basic component of a well-rounded educational program. In this chapter, we begin by introducing some of the terminology of data analysis and consider the nature and role of variability in statistical analyses.

1.1 Three Reasons to Study Statistics

Because of the widespread use of statistical analysis to organize, summarize, and draw conclusions from data, it is clear that a familiarity with statistical techniques and statistical literacy in general is vital in today's society. Here are three important reasons why it's a good idea for everyone to have a basic understanding of statistics, and why many college majors require a course in statistics.

The First Reason: Being an Informed "Information Consumer"

In today's society, we are bombarded with numerical information in news, in ads, and even in conversation. How do we decide whether claims based on numerical information are reasonable? Here are just a few examples from one day's news (*Los Angeles Times* and *San Francisco Examiner,* May 28, 1995):

■ *A statistical analysis of U.S. public high school reading and math scores.* This analysis showed that, in spite of headlines bemoaning the decline in SAT scores, scores in both reading and math have gone up. The study, conducted by the Rand Corporation, pointed out that SAT scores measure the achievement of college-bound students, who make up less than half of all high school students. U.S. Department of Education Achievement (USDEA) Tests, however, are taken by all students, not just those preparing for college. Analysis of USDEA test scores showed that both reading and math scores were higher in 1995 than in 1970, with the greatest gains reported for blacks and Latinos.

■ *An assessment of the impact of affirmative action on civil service workforce composition in California.* Comparisons based on ethnicity and gender were made using data from 1978 and from 1994. In a separate article, data from a public opinion poll on attitudes toward affirmative action programs was used to support a political candidate's claim that opposition to such programs is growing.

■ *An analysis of change in the average length of nine-inning major-league baseball games.* A graph in the article showed an increase from an average length of 2 hours and 33 minutes in 1981 to an average of 2 hours and 58 minutes in 1994. The analysis prompted some suggestions for shortening game times.

■ *A report on a marketing study that examined the customer base for check-cashing services* (businesses that charge a fee for cashing checks). The study presented data on the reasons people use this type of service rather than a bank, as well as data on client demographics, including race, gender, and age. The reported information is useful in evaluating the potential of a proposed location for such a business.

■ *An annual report on housing prices across the United States.* The average cost of a 2200-square-foot home for 280 cities ranged from a low of $92,125 in Fort Worth, Texas, to a high of $886,000 in Beverly Hills, California. An index for each city that compared the city to the national average also appeared. The article explained how to use these indexes to calculate the cost of replacing a house in one city with a comparable house in another.

To be an informed consumer of such reports, you must be able to do the following: (1) extract information from charts and graphs, (2) follow numerical arguments, and (3) know the basics of how data should be gathered, summarized, and analyzed to draw statistical conclusions.

The Second Reason: Understanding and Making Decisions

No matter what profession you choose, you will almost certainly need to understand statistical information and base decisions on it. Here are some examples:

■ Almost all industries, as well as government and nonprofit organizations, use market research tools, such as consumer surveys, designed to reveal who uses their products or services.

■ Modern science and its applied fields, from astrophysics to zoology, rely on statistical methods for analyzing data and deciding whether various conjectures are supported by observed data. This is true for the social sciences, such as economics and psychology, and increasingly, liberal arts fields, such as literature and history, use statistics as a research tool.

■ In law or government, you may be called on to understand and debate statistical techniques used in another field. Class-action lawsuits can depend on a statistical analysis of whether one kind of injury or illness is more common in a particular group than in the general population. Proof of guilt in a criminal case may rest on statistical interpretation of the likelihood that DNA samples match.

Throughout your professional life, you will have to make informed decisions and assess the risk of various choices. To make these decisions, you must be able to do the following:

1. Decide whether existing information is adequate or whether additional information is required.
2. If necessary, collect more information in a reasonable and thoughtful way.
3. Summarize the available data in a useful and informative manner.
4. Analyze the available data.
5. Draw conclusions, make decisions, and assess the risk of an incorrect decision.

Statistical methods are the tools for accomplishing these steps. You may already informally use these steps to make everyday decisions. Should you go out for a sport that involves the risk of injury? Will your college organization do better by trying to raise funds with a benefit concert or a direct appeal for donations? If you choose a particular major, what are your chances of finding a job when you graduate? How should you select a graduate program based on guidebook ratings that include information on percentage of applicants accepted, time to obtain a degree, and so on?

The Third Reason: Evaluating Decisions That Affect Your Life

An understanding of statistical techniques will allow you to evaluate decisions that affect your well-being. Here are some instances:

■ Insurance companies use statistical techniques to set auto insurance rates, although some states restrict the use of these techniques. Data suggests that young drivers have more accidents than older ones. Should laws or regulations limit how much more the young drivers pay for insurance? What about the common practice of charging higher rates for people who live in inner cities?

■ University financial aid offices survey students on the cost of going to school and collect data on family income, savings, and expenses. The resulting data is used to set criteria used in deciding who receives financial aid. Are the estimates accurate?

■ Medical researchers use statistical methods to make recommendations regarding the choice between surgical and nonsurgical treatment of diseases such as coronary heart disease and cancer. How do they weigh the risks and benefits to reach such a recommendation?

■ Many companies now require drug screening as a condition of employment, but with these screening tests there is a risk of a false positive reading (incorrectly indicating drug use) or a false negative reading (failure to detect drug use). What are the consequences of a false result? Given the consequences, is the risk of a false result acceptable?

An understanding of elementary statistical methods can help you to decide whether important decisions like the ones just mentioned are being made in a reasonable way.

We encounter data and conclusions based on data every day. **Statistics** is the scientific discipline that provides methods to help us make sense of data. Some people regard conclusions based on statistical analyses with a great deal of suspicion. Extreme skeptics, usually speaking out of ignorance, characterize the discipline as a subcategory of lying — something used for deception rather than for positive ends. However, we believe that statistical methods, used intelligently, offer a set of powerful tools for gaining insight into the world around us. We hope that this text will help you to understand the logic behind statistical reasoning, prepare you to apply statistical methods appropriately, and enable you to recognize when others are not doing so.

1.2   Statistics and Data Analysis

Statistical methods, used appropriately, allow us to draw conclusions based on data. Data and conclusions based on data appear regularly in a variety of settings: newspapers, advertisements, magazines, and professional journals. In business, industry, and government, informed decisions are often data-driven.

Statistics is the science of collecting, analyzing, and drawing conclusions from data.

Once data has been collected or an appropriate data source identified, the next step in the data analysis process usually involves organizing and summarizing the information. Tables, graphs, and numerical summaries allow increased understanding and provide an effective way to present data. Methods for summarizing data make up the branch of statistics called **descriptive statistics.**

After data has been summarized, we often wish to draw conclusions or make decisions based on the data. This usually involves generalizing from a small group of individuals or objects that we have studied to a much larger group.

> **Definition**
>
> The entire collection of individuals or objects about which information is desired is called the **population** of interest. A **sample** is a subset of the population, selected for study in some prescribed manner.

For example, the admissions director at a large university might be interested in learning why some applicants who were accepted for the fall 2000 term failed to enroll at the university. The population of interest to the director consists of all accepted applicants who did not enroll in the fall 2000 term. Because this population is large and it may be difficult to contact all the individuals, she might be able to collect data from only 300 selected students. These 300 students constitute what is known as a sample.

The second major branch of statistics, **inferential statistics,** involves generalizing from a sample to the population from which it was selected. When we generalize in this way, we run the risk of an incorrect conclusion, since a conclusion about the population will be reached on the basis of incomplete information. An important aspect in the development of inferential techniques involves quantifying the chance of an incorrect conclusion.

Considering some examples will help you develop a preliminary appreciation for the scope and power of statistical methodology. In the following examples, we describe three problems that can be investigated using techniques to be presented in this text.

EXAMPLE 1.1

Suppose that a university has recently implemented a new phone registration system. By entering information from a touch-tone phone, students interact with the computer to select classes for the term. To assess student opinion regarding the effectiveness of the system, a survey of students is to be undertaken. Each student in a sample of 400 will be asked a variety of questions (such as the number of units received, the number of attempts required to get a phone connection, etc.). The survey will yield a rather large and unwieldy data set. To make sense out of the raw data and to describe student responses, it is desirable to summarize the data. This would also make the results more accessible to others. Descriptive techniques can be used to accomplish this task. In addition, inferential methods can be employed to draw various conclusions about the experiences of *all* students who used the registration system.

EXAMPLE 1.2

A study linking depression to low cholesterol levels is described in an Associated Press article (*San Luis Obispo Telegram Tribune,* June 23, 1995). Researchers at a hospital in Italy compared the average cholesterol level for a sample of 331 patients who had been admitted to the hospital after a suicide attempt and diagnosed with clinical depression to the average cholesterol level of 331 patients admitted to the hospital for other reasons. Statistical techniques were used to analyze the data and to show that the average cholesterol level was lower for the depressed group. The article correctly noted that, because of the way in which the data was collected, it was not possible to determine from the statistical analysis alone whether there is a causal relationship between cholesterol level and psychological state — that is, whether low cholesterol levels affect psychological state or vice versa.

EXAMPLE 1.3   A final example comes from the discipline of forestry. When a fire occurs in a forested area, decisions must be made as to the best way to combat the fire. One possibility is to try to contain the fire by building a fire line. If building a fire line requires 4 hours, deciding where the line should be built involves making a prediction of how far the fire will spread during this period. Many factors must be taken into account, including wind speed, temperature, humidity, and time elapsed since the last rainfall. Statistical techniques make it possible to develop a model for the prediction of fire spread, using information available from past fires.

Exercises 1.1 – 1.7

1.1 Give a brief definition of the terms *descriptive statistics* and *inferential statistics.*

1.2 Give a brief definition of the terms *population* and *sample.*

1.3 The student senate at a university with 15,000 students is interested in the proportion of students who favor a change in the grading system to allow for + and − grades (e.g., B−, B, B+, rather than just B). Two hundred students are interviewed to determine their attitude toward this proposed change. What is the population of interest? What group of students constitutes the sample in this problem?

1.4 The supervisors of a rural county are interested in the proportion of property owners who support the construction of a sewer system. Because it is too costly to contact all 7000 property owners, a survey of 500 (selected at random) is undertaken. Describe the population and sample for this problem.

1.5 Representatives of the insurance industry wished to investigate the monetary loss due to earthquake damage to single-family dwellings in Northridge, California, in January of 1994. From the set of all single-family homes in Northridge, 100 homes were selected for inspection. Describe the population and sample for this problem.

1.6 A consumer group conducts crash tests of new-model cars. To determine the severity of damage to 1999 Mazda 626s resulting from a 10-mph crash into a concrete wall, six cars of this type are tested, and the amount of damage is assessed. Describe the population and sample for this problem.

1.7 A building contractor has a chance to buy an odd lot of 5000 used bricks at an auction. She is interested in determining the proportion of bricks in the lot that are cracked and therefore unusable for her current project, but she does not have enough time to inspect all 5000 bricks. Instead, she checks 100 bricks to determine whether each is cracked. Describe the population and sample for this problem.

1.3   The Nature and Role of Variability

Statistics is the science of collecting, analyzing, and drawing conclusions from data. If we lived in a world where all measurements were identical for every individual, all three of these tasks would be simple. Imagine a population consisting of all students at a particular university. Suppose that *every* student is taking exactly the same number of units, spent exactly the same amount of money on textbooks this semester, and favors increasing student fees to support expanding library services (this is fiction!). For this population, there is *no* variability in the values of number of units, amount spent on books, or student opinion on the fee increase.

A researcher studying this population to draw conclusions about these three variables would have a particularly easy task. It would not matter how many students the researcher included in his sample or how the sampled students were selected. In fact, the researcher could collect information on number of units, amount spent on books, and opinion on the fee increase by just stopping the next student who happened to walk by the library. Since there is no variability in the population, this one individual would provide complete and accurate information about the population, and the researcher could draw conclusions based on the sample with no risk of error.

The situation just described is obviously unrealistic. Populations with no variability don't exist — or if they do, they are of little statistical interest because they present no challenge! In fact, variability is almost universal. It is variability that makes life (and the life of a statistician, in particular) interesting. Given the presence of variability, how we collect data and how we analyze and draw conclusions from it require that we understand the nature of variability. One of the primary uses of descriptive statistical methods is to increase our understanding of the nature of variability in a population.

The following two examples illustrate how an understanding of variability is necessary to draw conclusions based on data.

EXAMPLE 1.4

The graphs in Figure 1.1 (page 8) are examples of a type of graph called a histogram. (We will see how to construct histograms in Chapter 3.) Figure 1.1(a) shows the distribution of the heights of female basketball players who played at a particular university between 1990 and 1998. The height of each bar in the graph indicates how many players' heights were in the corresponding interval. For example, 40 basketball players had heights between 72 in. and 74 in., whereas only 2 had heights between 66 in. and 68 in. Figure 1.1(b) shows the distribution of heights for members of the women's gymnastics team over the same period. Both histograms are based on the heights of 100 women.

The first histogram shows that the heights of female basketball players varied, with most heights falling between 68 in. and 76 in. Heights of female gymnasts also varied, with most heights in the range of 60 in. to 72 in. We can also see that there is more variation in the heights of the gymnasts than in the heights of the basketball players, since the heights of gymnasts tended to differ more from one another.

FIGURE 1.1 Heights of female athletes: (a) Basketball players; (b) Gymnasts

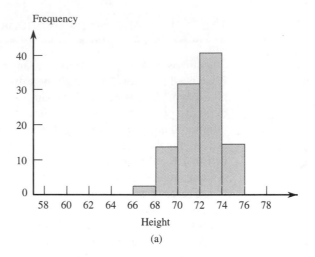

(a)

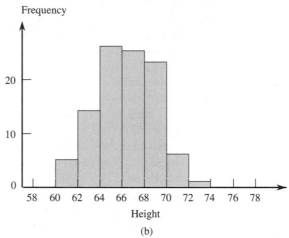

(b)

Now suppose that a tall woman (5 ft 11 in) tells you she is looking for her sister who is practicing with her team at the gym. Would you direct her to where the basketball team is practicing or to where the gymnastics team is practicing? What reasoning would you use to decide? What if you found a pair of size 6 tennis shoes left in the locker room? Would you first try to return them by checking with members of the basketball team or the gymnastics team?

You probably answered that you would send the woman looking for her sister to the basketball practice and try to return the shoes to a gymnastics team member. To reach these conclusions, you informally used statistical reasoning that combined your own knowledge of the relationship between heights of siblings and between shoe size and height with the information about the variability in heights presented in Figure 1.1. You might have reasoned that heights of siblings tend to be similar and that a height as great as 5 ft 11 in, although not impossible, would be unusual for a gymnast. On the other hand, a height as tall as 5 ft 11 in would be a common occurrence for a basketball player. Similarly, you might have reasoned that tall people tend to have bigger feet and short people tend to have smaller feet. The shoes found were a small size, so it is more likely that they belong to a gymnast than to a basketball player, since small heights and small feet are usual for gymnasts and unusual for basketball players.

EXAMPLE 1.5

As part of its regular water quality monitoring efforts, an environmental control board selects five water specimens from a particular well each day. The concentration of contaminants in parts per million (ppm) is measured for each of the five specimens, and then the average of the five measurements is calculated. The histogram in Figure 1.2 summarizes the average contamination values for 200 days.

Suppose a chemical spill has occurred at a manufacturing plant about 1 mile from the well. It is not known whether a spill of this nature would contaminate ground water in the area of the spill and, if so, whether a spill this distance from the well would affect the quality of well water.

One month after the spill, five water specimens are collected from the well and the average contamination is 16 ppm. Would you take this as convincing evidence that the well water was affected by the spill? What if the calculated average was 18 ppm? 22 ppm? How is your reasoning related to the graph in Figure 1.2?

Before the spill, the average contaminant concentration varied from day to day. An average of 16 ppm would not have been an unusual value, and so seeing an average of 16 after the spill isn't necessarily an indication that contamination has increased. On the other hand, an average as large as 18 ppm is less common, and an average as large as 22 is not at all typical of the prespill values. In this case, we would probably conclude that the well contamination level has increased.

FIGURE 1.2 Contaminant concentration in well water

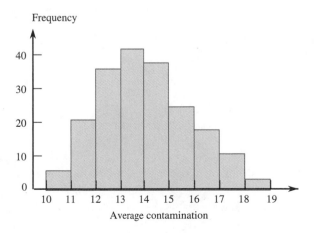

In each of the previous examples, reaching a conclusion required an understanding of variability. Understanding variability allows us to distinguish between usual and unusual values. The ability to recognize unusual values in the presence of variability is the essence of most statistical procedures and is also what enables us to quantify the chance of being incorrect when a conclusion is based on sample data. These concepts will be developed further in subsequent chapters.

Summary of Key Concepts and Formulas

Term or Formula	Comment
Descriptive statistics	Numerical, graphical, and tabular methods for organizing and summarizing data.
Population	The entire collection of individuals or measurements about which information is desired.
Sample	A part of the population selected for study.

References

Moore, David. *Statistics: Concepts and Controversies,* 4th ed. New York: W. H. Freeman, 1997. (A very nice, informal survey of statistical concepts and reasoning.)

Tanur, Judith, ed. *Statistics: A Guide to the Unknown.* Belmont, CA: Duxbury Press, 1989. (Short articles by a number of well-known statisticians and users of statistics, all very nontechnical, on the application of statistics in various disciplines and subject areas.)

2

The Data Analysis Process
and Collecting Data Sensibly

INTRODUCTION

An early step in the data analysis process is data collection. This step is critical because the type of analysis that is appropriate and the conclusions that can be drawn from it depend on how the data is collected. After describing the data analysis process in some detail, this chapter goes on to discuss two methods of data collection — sampling and experimentation.

2.1 Types of Data

Every discipline has its own particular way of using common words, and statistics is no exception. You will recognize some of the terminology from previous math and science courses, but much of the language of statistics will be new to you. We begin by introducing some common terms that will be used throughout this book.

The individuals or objects in any particular population typically possess many characteristics that might be studied. Consider as an example a group of students currently enrolled in a statistics course. One characteristic of the students in the population is the brand of calculator owned (Casio, Hewlett-Packard, Sharp, Texas Instruments, and so on). Another characteristic is the number of textbooks purchased, and yet another is the distance from the university to each student's permanent residence. A **variable** is any characteristic whose value may change from one individual to another. For example, *calculator brand* is a variable, and so are *number of textbooks purchased* and *distance to the university.* **Data** results from making observations either on a single variable or simultaneously on two or more variables. A **univariate data set** consists of observations on a single variable made on individuals in a sample or population.

In the previous example, *calculator brand* is a categorical variable, since each student's response to the query, "What brand of calculator do you own?" is a category. The collection of responses from all these students forms a **categorical data set.** The other two attributes, *number of textbooks purchased* and *distance,* are both numerical in nature. Determining the value of such a numerical variable (by counting or measuring) for each student results in a **numerical data set.**

Definition

A data set consisting of observations on a single attribute is a **univariate data set.** A univariate data set is **categorical** (or **qualitative**) if the individual observations are categorical responses; it is **numerical** (or **quantitative**) if each observation is a number.

EXAMPLE 2.1

The article "Knee Injuries in Women Collegiate Rugby Players" (*Amer. J. of Sports Medicine,* 1997: 360–362) reported the following data on type of injury sustained by 13 female rugby players. MCL, ACL, and the meniscus are different sites in the knee and the patella is the kneecap. (This is a subset of the data given in the article.)

meniscus tear	patella dislocation	MCL tear	MCL tear
meniscus tear	ACL tear	meniscus tear	meniscus tear
MCL tear	meniscus tear	ACL tear	patella dislocation
MCL tear			

Since type of injury is a categorical (nonnumerical) response, this is a categorical data set.

EXAMPLE 2.2

A sample of 20 compact cars, all of the same model, is selected, and the fuel efficiency (miles per gallon) is determined for each one. The resulting numerical data set is

29.8	28.5	27.6	29.5	28.3	27.2	28.7	26.9	27.9	28.4
30.1	28.0	28.0	30.0	28.7	29.6	27.9	29.1	29.9	27.9

In both of the preceding examples, the data sets consisted of observations (categorical responses or numbers) on a single variable, and so they are univariate data sets.

In some studies, attention focuses simultaneously on two different attributes. For example, both the height (in.) and weight (lb) might be recorded for each individual in a group. The resulting data set consists of pairs of numbers, such as

collected. Even if a decision is made to use existing data, it is important to understand how the data was collected and for what purpose, so that any resulting limitations are also fully understood and judged to be acceptable. If new data is to be collected, a careful plan must be developed, since the type of analysis that is appropriate and the conclusions that can be drawn from it are dependent on how the data is collected.

Data Summarization and Preliminary Analysis

After data is collected, the next step usually involves a preliminary analysis that includes summarizing the data graphically and numerically. This type of analysis provides insight into important characteristics of the data and can provide guidance in selecting appropriate methods for analysis.

Formal Data Analysis

The data analysis step requires the selection and application of appropriate statistical methods. Much of this text is devoted to methods that may be used to carry out this step.

Interpretation of Results

A critical step in the data analysis process is the interpretation of results. The purpose of this step is to address the following questions:

- What conclusions can be drawn from the analysis?
- How do the results of the analysis inform us about the stated research problem or question?
- How can our results guide future research?

This step in the process often leads to the formulation of new research questions, which, in turn, leads us back to the first step. In this way, good data analysis is often an iterative process.

Evaluating a Research Study

The data analysis steps just described can also be used as a guide for evaluating published research studies. The steps suggest asking the following questions as part of the evaluation of a study:

- What were the researchers trying to learn? What questions motivated their research?
- Was relevant information collected? Were the right things measured?
- Was the data collected in a sensible way?
- Was the data summarized in an appropriate way?
- Was an appropriate method of analysis selected, given the type of data and how the data was collected?
- Are the conclusions drawn by the researchers supported by the data analysis?

(68, 146). This is called a **bivariate data set. Multivariate data** results from obtaining a category or value for each of two or more attributes (so bivariate data is a special case of multivariate data). For example, multivariate data would result from determining height, weight, pulse rate, and systolic blood pressure for each individual in a group. Much of this book will focus on methods for analyzing univariate data. In the last several chapters, we consider some methods for analyzing bivariate and multivariate data.

Two Types of Numerical Data

With numerical data, it is useful to make a further distinction between *discrete* and *continuous* numerical data. Visualize a number line (Figure 2.1) for locating values of the numerical variable being studied. To every possible number (2, 3.125, −8.12976, etc.) there corresponds exactly one point on the number line. Now suppose that the variable of interest is the number of cylinders of an automobile engine. The possible values of 4, 6, and 8 are identified in Figure 2.2(a) by the dots at the points marked 4, 6, and 8. These possible values are isolated from one another on the line; around any possible value, we can place an interval that is small enough that no other possible value is included in the interval. On the other hand, the line segment in Figure 2.2(b) identifies a plausible set of possible values for the time it takes a car to travel a quarter mile. Here the possible values comprise an entire interval on the number line, and no possible value is isolated from the other possible values.

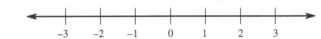

FIGURE 2.1 A number line

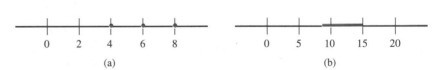

(a)   (b)

FIGURE 2.2 Possible values of a variable: (a) Number of cylinders; (b) Quarter-mile time

> **Definition**
>
> Numerical data is **discrete** if the possible values are isolated points on the number line. Numerical data is **continuous** if the set of possible values forms an entire interval on the number line.

Discrete data usually arises when each observation is determined by counting (the number of classes for which a student is registered, the number of petals on a certain type of flower, and so on).

EXAMPLE 2.3

The number of telephone calls per day to a drug hotline is recorded for 12 days. The resulting data set is

3 0 4 3 1 0 6 2 0 0 1 2

Possible values for the *number of calls* are 0, 1, 2, 3, . . .; these are isolated points on the number line, so we have a sample consisting of discrete numerical data.

The sample of fuel efficiencies in Example 2.2 is an example of continuous data. A car's fuel efficiency could be 27.0, 27.13, 27.12796, or any other value in an entire interval. Other examples of continuous data are task completion times, body temperatures, and package weights. In general, data is continuous when observations involve making measurements, as opposed to counting.

In practice, measuring instruments do not have infinite accuracy, so possible measured values, strictly speaking, do not form a continuum on the number line. However, any number in the continuum *could* be a value of the variable. The distinction between discrete and continuous data will be important in our discussion of probability models.

Exercises 2.1 – 2.4

2.1 Classify each of the following attributes as either categorical or numerical. For those that are numerical, determine whether they are discrete or continuous.
a. Number of students in a class of 35 who turn in a term paper before the due date
b. Gender of the next baby born at a particular hospital
c. Amount of fluid (oz) dispensed by a machine used to fill bottles with soda pop
d. Thickness of the gelatin coating of a vitamin E capsule
e. Birth classification (only child, firstborn, middle child, lastborn) of a math major

2.2 Classify each of the following attributes as either categorical or numerical. For those that are numerical, determine whether they are discrete or continuous.
a. Brand of computer purchased by a customer
b. State of birth for someone born in the United States
c. Price of a textbook
d. Concentration of a contaminant (micrograms/cm^3) in a water sample
e. Zip code (Think carefully about this one.)
f. Actual weight of coffee in a 1-lb can

2.3 For the following numerical attributes, state whether each is discrete or continuous.

a. The number of insufficient-fund checks received by a grocery store during a given month
b. The amount by which a 1-lb package of ground beef decreases in weight (because of moisture loss) before purchase
c. The number of New York Yankees during a given year who will not play for the Yankees the next year
d. The number of students in a class of 35 who have purchased a used copy of the textbook
e. The length of a 1-year-old rattlesnake
f. The altitude of a location in California selected randomly by throwing a dart at a map of the state
g. The distance from the left edge at which a 12-in. plastic ruler snaps when bent sufficiently to break
h. The price per gallon paid by the next customer to buy gas at a particular station

2.4 For each of the following situations, give some possible data values that might arise from making the observations described.
a. The manufacturer for each of the next ten automobiles to pass through a given intersection is noted.
b. The grade point average for each of the 15 seniors in a statistics class is determined.
c. The number of gas pumps in use at each of 20 gas stations at a particular time is determined.

d. The actual net weight of each of 12 bags of fertilizer having a labeled weight of 50 lb is determined.
e. Fifteen different radio stations are monitored during a 1-hr period, and the amount of time devoted to commercials is determined for each.

f. The bran 16 customer
g. The num of the next 2 certain high

2.2 The Data Analysis Process

Statistics involves the collection and analy data without analysis is of little value, and tract meaningful information from data tl In this section, we give an overview of the framework for the material covered in this

Planning and Conducting a Study

Most scientific studies are undertaken to ar flu vaccine effective in preventing illness? Are injuries that result from bicycle accid mets than for those who do not? How man Do engineering students or psychology stu lection and analysis allow researchers to a

The data analysis process can be organ nature of the problem; (2) deciding what t lecting the data; (4) summarizing the data a mally analyzing the data; and (6) interpret nal problem.

Understanding the Nature of the Problem

Effective data analysis begins with an unc must know the goal of the researcher and important to have clear direction before g able to answer the questions of interest u

Deciding What to Measure and How to Measure

The next step in the process is deciding v questions of interest. In some cases, the c of the relationship between the weight of played), but in other cases it is not as strai relationship between preferred learning carefully define the variables to be studi determining these values.

Data Collection

The data collection step in this process i decide whether an existing data source i

(68, 146). This is called a **bivariate data set. Multivariate data** results from obtaining a category or value for each of two or more attributes (so bivariate data is a special case of multivariate data). For example, multivariate data would result from determining height, weight, pulse rate, and systolic blood pressure for each individual in a group. Much of this book will focus on methods for analyzing univariate data. In the last several chapters, we consider some methods for analyzing bivariate and multivariate data.

Two Types of Numerical Data

With numerical data, it is useful to make a further distinction between *discrete* and *continuous* numerical data. Visualize a number line (Figure 2.1) for locating values of the numerical variable being studied. To every possible number (2, 3.125, −8.12976, etc.) there corresponds exactly one point on the number line. Now suppose that the variable of interest is the number of cylinders of an automobile engine. The possible values of 4, 6, and 8 are identified in Figure 2.2(a) by the dots at the points marked 4, 6, and 8. These possible values are isolated from one another on the line; around any possible value, we can place an interval that is small enough that no other possible value is included in the interval. On the other hand, the line segment in Figure 2.2(b) identifies a plausible set of possible values for the time it takes a car to travel a quarter mile. Here the possible values comprise an entire interval on the number line, and no possible value is isolated from the other possible values.

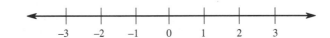

FIGURE 2.1 A number line

FIGURE 2.2 Possible values of a variable: (a) Number of cylinders; (b) Quarter-mile time

---

Definition

Numerical data is **discrete** if the possible values are isolated points on the number line. Numerical data is **continuous** if the set of possible values forms an entire interval on the number line.

Discrete data usually arises when each observation is determined by counting (the number of classes for which a student is registered, the number of petals on a certain type of flower, and so on).

EXAMPLE 2.3

The number of telephone calls per day to a drug hotline is recorded for 12 days. The resulting data set is

3 0 4 3 1 0 6 2 0 0 1 2

Possible values for the *number of calls* are 0, 1, 2, 3, . . .; these are isolated points on the number line, so we have a sample consisting of discrete numerical data.

The sample of fuel efficiencies in Example 2.2 is an example of continuous data. A car's fuel efficiency could be 27.0, 27.13, 27.12796, or any other value in an entire interval. Other examples of continuous data are task completion times, body temperatures, and package weights. In general, data is continuous when observations involve making measurements, as opposed to counting.

In practice, measuring instruments do not have infinite accuracy, so possible measured values, strictly speaking, do not form a continuum on the number line. However, any number in the continuum *could* be a value of the variable. The distinction between discrete and continuous data will be important in our discussion of probability models.

Exercises 2.1 – 2.4

2.1 Classify each of the following attributes as either categorical or numerical. For those that are numerical, determine whether they are discrete or continuous.
   a. Number of students in a class of 35 who turn in a term paper before the due date
   b. Gender of the next baby born at a particular hospital
   c. Amount of fluid (oz) dispensed by a machine used to fill bottles with soda pop
   d. Thickness of the gelatin coating of a vitamin E capsule
   e. Birth classification (only child, firstborn, middle child, lastborn) of a math major

2.2 Classify each of the following attributes as either categorical or numerical. For those that are numerical, determine whether they are discrete or continuous.
   a. Brand of computer purchased by a customer
   b. State of birth for someone born in the United States
   c. Price of a textbook
   d. Concentration of a contaminant (micrograms/cm^3) in a water sample
   e. Zip code (Think carefully about this one.)
   f. Actual weight of coffee in a 1-lb can

2.3 For the following numerical attributes, state whether each is discrete or continuous.

   a. The number of insufficient-fund checks received by a grocery store during a given month
   b. The amount by which a 1-lb package of ground beef decreases in weight (because of moisture loss) before purchase
   c. The number of New York Yankees during a given year who will not play for the Yankees the next year
   d. The number of students in a class of 35 who have purchased a used copy of the textbook
   e. The length of a 1-year-old rattlesnake
   f. The altitude of a location in California selected randomly by throwing a dart at a map of the state
   g. The distance from the left edge at which a 12-in. plastic ruler snaps when bent sufficiently to break
   h. The price per gallon paid by the next customer to buy gas at a particular station

2.4 For each of the following situations, give some possible data values that might arise from making the observations described.
   a. The manufacturer for each of the next ten automobiles to pass through a given intersection is noted.
   b. The grade point average for each of the 15 seniors in a statistics class is determined.
   c. The number of gas pumps in use at each of 20 gas stations at a particular time is determined.

d. The actual net weight of each of 12 bags of fertilizer having a labeled weight of 50 lb is determined.

e. Fifteen different radio stations are monitored during a 1-hr period, and the amount of time devoted to commercials is determined for each.

f. The brand of breakfast cereal purchased by each of 16 customers is noted.

g. The number of defective tires is determined for each of the next 20 automobiles stopped for speeding on a certain highway.

2.2 The Data Analysis Process

Statistics involves the collection and analysis of data. Both tasks are critical. Raw data without analysis is of little value, and even a sophisticated analysis cannot extract meaningful information from data that was not collected in a sensible way. In this section, we give an overview of the data analysis process, thus providing a framework for the material covered in this text.

Planning and Conducting a Study

Most scientific studies are undertaken to answer questions about our world. Is a new flu vaccine effective in preventing illness? Is the use of bicycle helmets on the rise? Are injuries that result from bicycle accidents less severe for riders who wear helmets than for those who do not? How many credit cards do college students carry? Do engineering students or psychology students pay more for textbooks? Data collection and analysis allow researchers to answer such questions.

The data analysis process can be organized into six steps: (1) understanding the nature of the problem; (2) deciding what to measure and how to measure it; (3) collecting the data; (4) summarizing the data and making a preliminary analysis; (5) formally analyzing the data; and (6) interpreting the results in the context of the original problem.

Understanding the Nature of the Problem

Effective data analysis begins with an understanding of the research problem. We must know the goal of the researcher and what questions are to be answered. It is important to have clear direction before gathering data in order to avoid being unable to answer the questions of interest using the data collected.

Deciding What to Measure and How to Measure It

The next step in the process is deciding what information is needed to answer the questions of interest. In some cases, the choice is obvious (for example, in a study of the relationship between the weight of a Division I football player and position played), but in other cases it is not as straightforward (for example, in a study of the relationship between preferred learning style and intelligence). It is important to carefully define the variables to be studied and to select an appropriate means of determining these values.

Data Collection

The data collection step in this process is a crucial one. The researcher must first decide whether an existing data source is adequate or whether new data must be collected. Even if a decision is made to use existing data, it is important to understand how the data was collected and for what purpose, so that any resulting limitations are also fully understood and judged to be acceptable. If new data is to be collected, a careful plan must be developed, since the type of analysis that is appropriate and the conclusions that can be drawn from it are dependent on how the data is collected.

Data Summarization and Preliminary Analysis

After data is collected, the next step usually involves a preliminary analysis that includes summarizing the data graphically and numerically. This type of analysis provides insight into important characteristics of the data and can provide guidance in selecting appropriate methods for analysis.

Formal Data Analysis

The data analysis step requires the selection and application of appropriate statistical methods. Much of this text is devoted to methods that may be used to carry out this step.

Interpretation of Results

A critical step in the data analysis process is the interpretation of results. The purpose of this step is to address the following questions:

■ What conclusions can be drawn from the analysis?

■ How do the results of the analysis inform us about the stated research problem or question?

■ How can our results guide future research?

This step in the process often leads to the formulation of new research questions, which, in turn, leads us back to the first step. In this way, good data analysis is often an iterative process.

Evaluating a Research Study

The data analysis steps just described can also be used as a guide for evaluating published research studies. The steps suggest asking the following questions as part of the evaluation of a study:

■ What were the researchers trying to learn? What questions motivated their research?

■ Was relevant information collected? Were the right things measured?

■ Was the data collected in a sensible way?

■ Was the data summarized in an appropriate way?

■ Was an appropriate method of analysis selected, given the type of data and how the data was collected?

■ Are the conclusions drawn by the researchers supported by the data analysis?